CONTENTS

Acknowledgements

This study has relied heavily on the copious *Proofs of Evidence* produced by the CEGB for the Sizewell 'B' Public Inquiry. We are grateful for additional assistance given by the CEGB in clarifying and extending their published evidence. Several people have been generous in providing background material and advice, including Alan Jones and Katherine Woolley of the Technical Change Centre, Professor K. Bhaskar of UEA and Richard Bending of the Cavendish Laboratory, Cambridge. Our methods follow closely those pioneered by Francis Cripps and Wynne Godley in their study of telecommunications investment, and we are grateful for additional advice given by Francis Cripps. We are also grateful for the considerable effort put into copy-editing this study by Ann Newton.

This research was commissioned by the National Union of Mineworkers. Responsibility for the views expressed in this study lies solely with the authors.

THE ECONOMIC CONSEQUENCES OF THE SIZEWELL 'B' NUCLEAR POWER STATION

STEPHEN FOTHERGILL, GRAHAM GUDGIN
and NIGEL MASON

University of Cambridge

Published by the Department of Applied Economics,
Sidgwick Avenue, Cambridge CB3 9DE

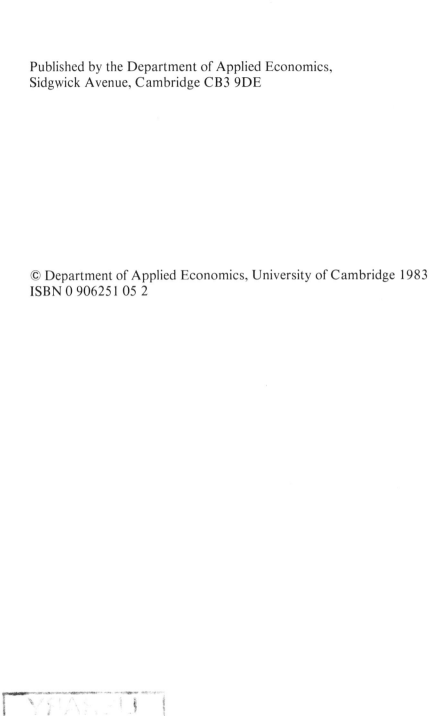

Printed in Great Britain by The Burlington Press (Cambridge) Ltd, Foxton, Cambridge CB2 6SW

Chapter 1

Introduction

The Central Electricity Generating Board (CEGB) proposes to build a nuclear power station, Sizewell 'B', on the Suffolk coast. The station will be the first in Britain to use a Pressurised Water Reactor (PWR). If the project secures approval the CEGB estimates that the capital cost will exceed £1.1 billion at March 1982 prices. Furthermore, the CEGB expects to order additional PWR stations during the 1980s and 1990s, in preference to all alternative types of generating plant, providing perhaps as much as 11 gigawatts (GW) of new capacity by the end of the century.[1] An investment in nuclear technology on this scale would probably be the largest investment programme in the history of the British economy.

This study investigates the economic implications of building and operating the Sizewell 'B' nuclear power station. The CEGB's economic evaluation has been undertaken along narrow commercial lines.[2] The CEGB attempts to measure the costs and savings which it would incur in proceeding with Sizewell 'B' and compares these with the costs and savings associated with alternative new generating capacity. The CEGB concludes that it is more economic, from its own point of view, to proceed immediately with the construction of Sizewell 'B' even though this will mean the premature retirement of existing coal-fired and oil-fired stations.

However, the construction and operation of a new power station has far wider economic consequences than those considered by the CEGB. Firstly, the decision to build a power station, and the choice of design, has major implications for those dependent on income from the CEGB and its suppliers. Secondly, taxpayers are affected by alterations in tax rates which the government may make in order to finance power station construction. Thirdly, since the construction and operation of a power station will alter the balance of payments, there will be adjustments in the level of income and employment in the economy as a whole. Each of these consequences is potentially large, and each is ignored in the CEGB analysis. Furthermore, it is by no means clear that the option which appears best from the CEGB's point of view is necessarily the best if these additional consequences are also considered. Indeed, from the point of view of the national economy the best use of limited funds might be to divert investment from the CEGB to more productive projects in other nationalised industries.

Sizewell 'B' should therefore be evaluated in terms of its impact on the economy as a whole; the economic case presented by the CEGB, which entirely ignores these wider issues, is an inadequate and partial basis for public decision-making. There is nothing new in the approach we are proposing. In Britain in recent years at least two major decisions by nationalised industries have been the subject of a macroeconomic evaluation: the choice of telephone exchange technology[3] and the closure of the Corby steelworks.[4] Each of these decisions was large enough to have a substantial impact on the whole economy. Sizewell 'B' is important for the same reason. The CEGB estimate of £1.1 billion for construction costs has already been noted. Allowing for interest during construction and possible cost over-runs, the total investment could well be in excess of £2 billion at March 1982 prices.

This study considers the consequences of building and operating Sizewell 'B' for the rest of the economy. Since details of alternative investments in other nationalised industries are not available it is not possible to compare Sizewell 'B' with other potential public sector investments, but the macroeconomic analysis we present provides a yardstick against which other investments can be assessed. The remainder of this introductory chapter deals briefly with the background to the current proposal for a PWR and the basis on which the CEGB justifies investment in nuclear power stations. The second chapter sets out the policy options now facing the government in determining the programme of investment in electricity supply. Chapter 3 describes the methodology we have adopted, and Chapters 4 and 5 describe the effect of each option on other industries and on the economy as a whole. Chapter 6 summarises the conclusions and presents policy recommendations.

The background to Sizewell 'B'

If it is built, Sizewell 'B' will be the first of a new generation of nuclear power stations in Britain. The first generation Magnox stations use a British design in which the reactor core is cooled by carbon dioxide gas, which in turn is used to raise steam to drive turbine generators. The stations were completed between 1962 and 1971 and are still in operation. All of them are small by comparison with most of the coal-fired and oil-fired stations built during this period. For several years the CEGB claimed that the Magnox stations generated cheaper electricity than coal-fired stations of the same vintage. However, the House of Commons Select Committee on Energy[5] and the Monopolies and Mergers Commission[6] both pointed out that these comparisons were misleading and based on defective accounting procedures which mixed historic and current costs. The marginal cost of operating the Magnox stations is lower than for coal-fired or oil-fired stations; this is true of all nuclear power stations. But reassessment of the total cost of electricity from the Magnox stations, basing capital charges on the current replacement cost of the stations rather than historic costs, has demonstrated that their electricity has been substantially more expensive than that from coal-fired stations of the same age.[7] As a source of cheap electricity the Magnox stations were not a 'good buy'.

The second generation nuclear power stations, the Advanced Gas-Cooled Reactors (AGRs), were confidently expected to produce cheaper electricity than any fossil-fuelled station. The AGR is a development of the Magnox reactor using basically the same method of cooling, but it operates at a higher temperature, uses a more enriched uranium oxide fuel, and was designed to generate larger quantities of electricity. Four AGR stations were ordered by the CEGB in the late 1960s. Their construction has been extremely protracted, to a large extent because of design problems with the reactor itself, and only one, Hinkley Point 'B', was completed before 1982. A further AGR station, Heysham II, was ordered by the CEGB in 1978 (plus another for the South of Scotland Electricity Board).

Electricity from Hinkley Point 'B' has so far proved no cheaper than electricity from the comparable coal-fired station Drax 'A'.[8] Of the three AGRs just finished or nearing completion, only one seems likely to produce cheaper electricity over its lifetime than Drax 'B' (a coal-fired station still under construction) even if the real increase in coal prices is 2.5% per annum compared to a 1% per annum real increase in nuclear fuel costs. At current fuel costs, AGRs will certainly not prove good investments.

The failure of the AGR programme to live up to expectations and the difficulty in achieving a settled design led to disillusion with the commitment to British reactors. In addition, no export orders were forthcoming. The result was the during the 1970s a number of industrial companies, notably GEC, lobbied for the adoption of the Pressurised Water Reactor as the basis for future British nuclear investment. The PWR is an American design which is used widely in the United States and in other countries and accounts for the majority of the world's installed nuclear capacity. It differs from British designs in that the reactor core is cooled by water under very high pressure, rather than by gas. The proposal to build a PWR at Sizewell reflects the acceptance by the CEGB of the view that the AGR no longer provides the best way forward for nuclear electricity. The Sizewell PWR will be based on the Standard Nuclear Unit Power Plant System (SNUPPS) designed by the Westinghouse Corporation, two examples of which are currently under construction in the United States.

Nevertheless, whatever the technical merits of alternative reactors, the CEGB's case for constructing Sizewell 'B' is still primarily economic. Attention needs to be drawn, however, to the exact nature of this economic case. The important point is that it is not based upon an anticipated shortage of generating capacity. Sizewell 'B' would be commissioned in 1992; the CEGB forecasts that it will probably not need Sizewell 'B' to make good any shortfall in generating capacity until 1996.[9] Given the CEGB assessment of future electricity demand there is therefore no need, on capacity grounds, to order Sizewell 'B' or any other new power station for at least another five years.

The CEGB argues that Sizewell 'B' should be ordered ahead of demand because it will provide cheaper electricity than all other alternatives. Other arguments are also deployed (e.g. the need to establish the PWR as a viable option for future investment) but the central economic argument for proceeding with the project is that it will lower the cost of electricity. Sizewell 'B' will be more economic, from the CEGB's point of view, than alternative new generating capacity and also than continuing to operate older power stations, mainly burning coal or oil, even though these older stations will therefore be retired before they reach the end of their design lives.

According to the CEGB, Sizewell 'B' will be cheaper to build than an equivalent AGR, though still substantially more expensive (in terms of £s per KW) than a coal-fired station. It will generate cheaper electricity because the high capital charges resulting from its construction will be more than offset by 'net system savings': Sizewell will use relatively cheap uranium fuel and lead to substantial savings of coal and oil which would otherwise be burned in older power stations. Given that the CEGB expects large increases in the real cost of fossil fuels, the savings anticipated by the CEGB become increasingly large during the lifetime of Sizewell 'B'.

Chapter 2

Policy Options

A number of options are available to the government in deciding the strategy the CEGB should follow during the next few years. They can allow the CEGB to build Sizewell 'B'. Alternatively, a further AGR station could be built. This would have the advantage that the reactor would be designed and built almost entirely in Britain but, in view of the disappointing record of the existing AGRs, the higher estimated capital costs and the CEGB's enthusiasm for the PWR, it is appropriate to regard the PWR as the principal nuclear contender. Additional oil-fired capacity remains a technical possibility but we share the CEGB's view that the price of oil no longer makes this a viable option. Similarly, renewable sources of energy – wind, waves and solar power – are not yet economically and technically viable according to the CEGB and we have not sought to challenge this view. The government could opt for a new coal-fired station or invest in a combined heat and power district heating scheme which used coal, or decide that no new power station of any type should be built until needed to meet the demand for electricity. Finally, investment in energy conservation is an alternative to investment in new generating capacity.

This study considers three of these options: Sizewell 'B'; a coal-fired power station; and no new building ahead of demand. These are the principal alternatives considered by the CEGB and ones which could be examined satisfactorily within the limits of this study. This chapter explains these options and the assumptions introduced about each. It also comments briefly about a possible programme of PWR construction.

Sizewell 'B'
The output from Sizewell 'B' would be 1110 MWso,[1] from one reactor driving two 660 MW turbine generators. Since the PWR design is new to Britain considerable uncertainty surrounds the construction costs and operating performance of Sizewell 'B'. The best guide is experience in the United States in building and operating the larger PWRs designed by Westinghouse. These have been the subject of studies by Komanoff,[2] and in several important respects his figures diverge from those presented by the CEGB.

Komanoff calculates that including interest during construction (at 5 per cent) the capital cost of a US-built Westinghouse PWR commissioned in 1988 will be $2180 per KW at mid 1981 prices, and that the two SNUPPS plants under construction are unlikely to diverge from this norm. The equivalent figure for Sizewell 'B', presented by the CEGB, is £1502 per KW at March 1982 prices.[3] At the current exchange rate (roughly £1 = $1.50) and adjusting for inflation the two estimates of capital costs do not differ greatly. However, Komanoff also calculates that a Westinghouse PWR commissioned in 1988 will cost 2.3 times more than the anticipated cost of a 1988-commissioned US coal plant. The CEGB, on the other hand, projects a capital cost ratio of only 1.6 for Sizewell 'B' *vis à vis* a coal-fired alternative. Moreover, the CEGB plans to incorporate additional safety features into the Sizewell PWR while British coal-fired power stations are built without some of the expensive pollution control equipment (notably sulphur 'scrubbers') included in American stations.

The CEGB's figures on the load factor[4] likely to be achieved by Sizewell 'B' are also open to question. The reason for this is that they are based on the recent operating performance of all Westinghouse PWRs over 1000 MW. However, the Electricity Consumers' Council argues that a more appropriate basis on which to assess the likely performance of Sizewell 'B' is to use a sample of all 4-loop Westinghouse reactors (the same design as Sizewell 'B') including two plants rated at less than 1000 MW. The performance of these reactors indicates a 'settled-down' load factor of 58 per cent.[5]

This evidence suggests that the CEGB may be optimistic in its forecasts of the cost and performance of the Sizewell PWR.[6] In addition, the large real cost escalation associated with the British AGR programme does not inspire confidence in CEGB estimates. In examining the economic consequences of building Sizewell 'B' it is therefore appropriate to consider two sets of costings: the figures presented by the CEGB to the Public Inquiry, and an alternative set which are less favourable to nuclear power.

The alternative assumptions we adopt are, firstly, that the capital costs of Sizewell 'B' will be 44 per cent higher in real terms[7] than those presented by CEGB. This brings the ratio between British PWR and coal station capital costs into line with the ratio in the United States. Secondly, we assume a 'settled-down' load factor of 58 per sent instead of the 64 per cent used by the CEGB. These alternative assumptions are not accurate predictions of what we expect to occur; rather, they are intended to be illustrative of a significant escalation of capital costs and a shortfall in operating performance.

A new coal-fired power station

During the 1960s and early 1970s the CEGB built several very large coal-fired stations of up to 1875 MW capacity incorporating combinations of 500 and 660 MW turbine generators. Drax 'B', due for completion in 1986, is the most recent station built to this basic design. As a result of the experience accumulated during the construction and operation of these stations the CEGB now has a well-defined design which would form the basis of any future order for a coal-fired station. This would comprise 3×660 MW turbine generators with a total capacity of 1875 MWso.

If the CEGB were not to receive permission to build a PWR at Sizewell it would not wish to proceed with the immediate construction of a coal-fired alternative. According to CEGB calculations the 'net effective cost' of building a new coal station is higher than the 'net avoidable cost' of keeping older stations in operation. This is partly because the costs of building coal stations have escalated in real terms, and partly because the overall fuel savings would be modest even though a new station would be more efficient than an older plant. A new coal station could not therefore be justified on cost minimisation grounds, and the CEGB would defer its construction until it was needed on capacity grounds. For comparative purposes, however, we contrast the economics of Sizewell 'B' with a coal-fired station commissioned at the same time (i.e. in 1992).

Because the CEGB has recent experience of building and running large coal-fired stations there are few grounds for questioning its central estimates of construction time and costs and operating performance. However, the CEGB's assessment of the economics of coal-fired stations is critically dependent on their assumptions concerning future coal prices, about which there is considerable uncertainty.

The central estimates used by the CEGB assume that the average pithead price of coal sold to the CEGB will rise by 1.7 per cent a year in real terms to 2000, and slightly faster thereafter.[8] This view is not based on a judgement that the real wages of miners will exert an upward influence on coal prices, but on the assumption that other operating costs,[9] which currently comprise half of coal production costs, will rise by 2.5 per cent a year until the end of the century. The CEGB also assumes that during the same period the world price of coal will rise and that by the mid 1990s the world price will be roughly the same as production costs in the UK. It is assumed that as British coal becomes internationally competitive the Exchequer subsidy to coal prices (currently about 10 per cent) will be removed, and after the mid 1990s the CEGB assumes that National Coal Board (NCB) prices will therefore reflect the rising world price rather than domestic production costs.

The very large rise in the real price of coal implied by these assumptions makes a new coal-fired power station increasingly costly. Conversely, it makes Sizewell 'B' appear an attractive option since the PWR will reduce the amount of expensive coal which the CEGB has to burn. Higher future coal prices than those assumed by the CEGB central estimates would not alter the CEGB's preference for PWR capacity; it is, however, vital to test the robustness of their case by examining the consequences of lower coal prices. In our evaluation of coal and nuclear power stations two sets of assumptions about coal prices are therefore used: one based on CEGB estimates and the other on a similar increase in the real price of coal.

The alternative assumption used is that coal prices will rise in real terms at 0.85 per cent a year. This is half the rate of increase anticipated by the CEGB, and roughly the same as the rate of increase between 1960 and 1973, when a slow rise in miners' real wages was combined with rapid increases in labour productivity due to mechanisation and the closure of less efficient pits. A similarly low rate of increase in the real price of coal in the future would probably require some combination of a continuing subsidy to coal and a large rise in miners' productivity. Alternatively, miners' wages would have to fall relative to other wages or the process of closing old pits and replacing them with newer more efficient mines would have to be accelerated. It is not possible to estimate the likelihood of these various possibilities, though there is already evidence that productivity in existing mines has begun to rise more rapidly and that the potential for further increases in efficiency is extensive.[10]

The no-new-station option

A third option open to the government is not to build any new power station during the next few years. The CEGB acknowledges that the immediate construction of Sizewell 'B' cannot be justified by a potential shortfall in generating capacity. After allowing for external supplies of electricity (e.g. *via* the cross-channel link), for the need to maintain a 28 per cent planning margin above maximum system demand, and for the retirement of older stations at the end of their nominal lives, existing and committed power stations will provide sufficient generating capacity until between 1993 and 2003, depending on the growth of the economy. Table 2.1 illustrates this point. Scenario C is the most relevant because it is the central forecast of national economic growth on which the CEGB case for Sizewell 'B' is based. On this scenario a shortage of capacity will not appear until 1996. In other words, given a 7.5-year construction period there is no need on capacity grounds to order Sizewell 'B' until 1988.

Table 2·1 The requirement for new generating capacity

	Average growth in GDP 1979/80–2000 % p.a.	'Shortage' of capacity appears in
Scenario A	2.6	1994
Scenario B	2.6	1993
Scenario C	1.0	1996
Scenario D	−0.4	2003
Scenario E	−0.4	2003

Source: CEGB, *Proof of Evidence, P4*, Tables 2 and 5.

We are in broad agreement with the CEGB about the growth of the economy to 1990 under Scenario C.[11] It implies an average growth of GDP of one per cent a year from 1979/80 to the end of the century. Table 2.2 compares Scenario C with forecasts of GDP in 1990 published in the *Cambridge Economic Policy Review*, vol. 8 no 1. These are the output of a macroeconomic model which has a consistently good record in forecasting medium-term trends in economic activity and unemployment. Much depends on the nature of economic policy over the next decade. If there is a radical change and reflationary policies are introduced, with import controls, by 1990 GDP will exceed that assumed by Scenario C. Reflation without import controls is predicted to lead to a level of GDP in 1990 similar to that assumed in Scenario C. If present policies are continued, the Cambridge model predicts only modest recovery, so that in 1990 GDP will still be far below the level assumed in Scenario C. The GDP growth assumed in Scenario C therefore falls roughly in the middle of these three possible outcomes and we have used it in our calculations, although we do not necessarily follow the details of Scenario C in relation to such variables as the level of unemployment, the rate of productivity growth and the exchange rate. After 1990 the growth of the economy cannot reasonably be predicted, though the assumption of one per cent a year growth in GDP, implied by Scenario C, is not implausible.

Table 2·2 GDP forecasts for 1990 (1979 = 100)

CEGB Scenario C	112
Present policies	103
Reflation beginning in 1984	114
Reflation *plus* import controls beginning in 1984	126

Sources: CEGB, *Proof of Evidence, P4*; *Cambridge Economic Policy Review*, Vol. 8 No. 1.

The problems inherent in forecasting are also indicated by the previous record of the electricity supply industry, which reveals a persistent tendency to overestimate electricity demand. The Electricity Council prepares annual forecasts of winter peak demand seven years ahead – roughly the period needed to build new power stations – which are adopted by the CEGB as the basis for planning investment in new capacity. Fig. 2.1 compares the forecasts with the outturn, and reveals that, throughout the last decade-and-a-half, actual demand has been much lower than forecast, with the consequence that in recent years the new capacity completed to meet optimistic forecasts of demand has simply led to the premature retirement of older power stations. Heysham II, the last of the AGRs, is due for completion in 1988 and continues this tradition: CEGB figures now show that it will not be needed on capacity grounds until 1995.[12]

CEGB estimates and evidence as to their plausibility therefore indicate that a realistic option open to the government is to postpone the constructions of any new power station. Moreover, even if there were a larger upturn in GDP and electricity demand than any current forecast anticipates, the CEGB would not face serious problems.

One reason for this is that because the future level of electricity demand is reviewed regularly the ordering of new power stations could be brought forward as soon as potential shortfall in capacity is identified, so that in the event it is unlikely that any actual shortfall would be large. A second reason is that the CEGB maintains a generous 28 per cent margin of spare capacity – the 'planning margin' – in excess of demand, to allow for breakdowns, variability in the weather and forecasting errors. The way in which a 'shortage' of capacity would be felt, therefore, would be by the planning margin falling below the 28 per cent target until new capacity came on stream. In this context it is worth noting that the planning margin was

Fig. 2·1 Electricity demand: forecast and outturn, 1956/7–1981/2

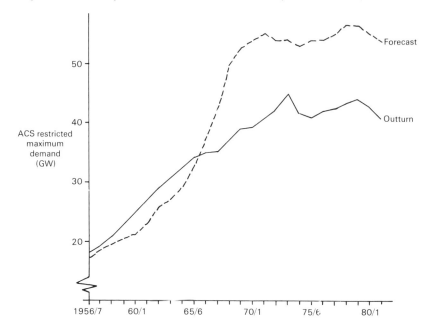

Sources: Select Committee on Energy 1980–81; CEGB, *Proof of Evidence, P4.*

previously much lower – 20 per cent between 1970 and 1976, and 17 per cent between 1964 and 1970 – but no consumers have been disconnected because of shortages of generating capacity since 1965–66.[13] The third reason why a shortfall in capacity need not cause problems is that the life of existing older plants could be prolonged to make good the deficiency. The notion of a 'shortfall' assumes the retirement of existing plant when it reaches the end of its 'nominal life' (40 years for coal-fired stations, for example). In practice some plants may need to be retired earlier for technical reasons, but it is probable that with limited refurbishment or replacement of components others could continue to operate beyond the end of their nominal lives to compensate temporarily for an unanticipated shortage of capacity, particularly as they would need to be used for only short periods to meet peak demand.

There are also some advantages to the CEGB in deferring the construction of power stations until they can be justified on capacity grounds. These arise because the technical and economic context within which investment in new plant occurs can change rapidly, making even quite new plant obsolete or unprofitable.[14] The best examples are the large oil-fired stations, which seemed sensible bets in the early 1970s but which now, as they reach completion, are too costly to use as base-load stations because of the rise in oil prices. Similar risks apply to the Sizewell PWR. A large shift in the relative prices of nuclear and fossil fuels, or a PWR accident elsewhere in the world necessitating expensive additional safety features at Sizewell,[15] could mean that Sizewell 'B' is uneconomic by the time it is needed. Deferring the construction of a new plant until it is justified on capacity grounds enables the decision on its design to be taken in the light of the technical and economic information available at that time – which may differ significantly from the information currently available.[16]

A PWR programme

Though the CEGB will not at present make a firm commitment it is their intention that Sizewell 'B' should be the first of a programme of PWRs in Britain. The impact of such a programme is not assessed in this study, but it is worth looking briefly at what it would involve and at its financial implications.

Fig. 2.2 shows the annual burden of financing the programme of PWR construction assumed by the CEGB for its 'high nuclear background'. This is the background against which the CEGB's central evaluation of Sizewell 'B' is conducted, and it represents the programme with which the CEGB would wish to proceed if there are no unforeseen changes of circumstances or problems with the PWR. It implies that by the end of the 1990s 11.1 GW of PWR capacity will be in operation, and 52.2 GW by 2030. The annual expenditure in Fig. 2.2 is calculated from the CEGB's estimates of PWR capital costs, the CEGB timetable of station commissionings,[17] and the

Fig. 2·2 A PWR and coal-fired programme compared

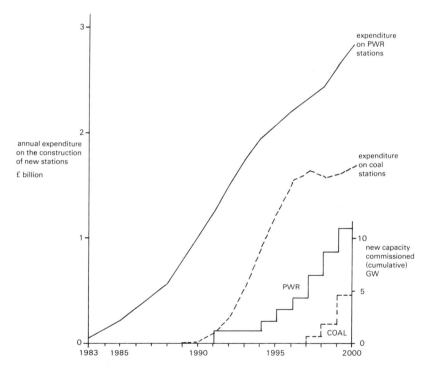

Source: Calculated from CEGB, *Proof of Evidence, P4*, Table 26C.

time-profile of capital expenditure on a PWR,[18] and a reduction has been made for savings of £90 million on National Nuclear Corporation (NNC) costs for PWRs subsequent to Sizewell 'B'.[19] Expenditure on a programme of coal-fired stations is also shown.[20] These would not be commissioned in advance of need, so the investment in coal stations in Fig. 2.2 represents the minimum necessary to provide capacity to meet the electricity demand forecast associated with Scenario C.

The striking feature is the scale of investment required by the PWR programme. Capital spending would rise every year until the end of the century, when it would reach a staggering £2.85 billion per year (at March 1982 prices). This is more than three times the total capital investment by the CEGB in 1981/2, and equals roughly half the combined annual capital expenditure of all nationalised industries at the present time. It also equals more than 40 per cent of the current level of fixed capital investment by the entire UK manufacturing sector. Indeed, if the programme implied by the CEGB's 'high nuclear background' goes ahead then by 2000 a cumulative total of £24.8 billion (at March 1982 prices) will have been spent on PWRs, and no fewer than 20 PWR stations will be at some stage of construction during that year.

Chapter 3

Economic Framework

In analysing the consequences for the economy of building and operating a new power station it is useful to distinguish between three effects: first, the direct effect on different industries; second, the effect on the Exchequer of financing a new power station; and third, the effect on the level of income and employment in the economy as a whole. This chapter describes how we have measured each of these effects and the assumptions it has been necessary to make about the way the British economy works.

Direct effects

The construction and operation of a power station involves a large volume of purchases from a wide range of firms both in the UK and abroad. These firms in turn buy components and materials from other firms, again both in the UK and abroad, and so on. Output and employment in all these firms are likely to respond to orders placed by the CEGB. In addition, the construction and operation of a new power station affects employment within the CEGB itself, not just at the new station but also at older stations which may be displaced by the new capacity. We describe these consequences collectively as the 'direct effects' of a new power station.[1]

The CEGB provides a breakdown of the estimated capital cost of different types of power station, enabling individual items of expenditure to be allocated to particular industries. The employment generated in those industries depends on the amount that is imported directly by the CEGB, the value-added per employee in the industries, and the ratio of bought-in components and materials to output. Bought-in components and materials generate further employment in other industries in the UK, the number depending again on the proportion of imports, value-added per employee and the amount of bought-in supplies. The methods used to derive these estimates are described in detail in the Appendix and have involved the use of current input-output tables of the UK economy together with assumptions about future levels of import penetration and value-added per employee.

Implicit in this approach is the view that orders placed by the CEGB, or its suppliers, lead to increases in production and employment in the industries concerned. An alternative view would be that CEGB orders simply displace other orders, so that there is no change in either production or employment in the industries in question. This latter view might be valid in an economy in which resources were already fully utilised. Our approach, which assumes an increase in orders leads to an increase in production, is justified by the likelihood of considerable spare capacity in the UK for the forseeable future.

Where the oil industry is concerned we have departed from this assumption. In their early days of operation, Sizewell 'B' or a coal-fired alternative would reduce the amount of oil burned in existing oil-fired stations. We assume that any oil not used by the CEGB would be exported, so there would be no change in production or employment in the UK oil industry.

We do not treat the coal industry in the same way as the oil industry. On CEGB estimates the world price of coal will be above NCB average production costs after the mid 1990s, so that the NCB might be able to export the coal released from electricity generation by Sizewell 'B'. However, if marginal changes in production affect the pits with the highest costs, then coal from those pits would not be internationally competitive unless cross-subsidisation occurred within the NCB. We assume that no such cross-subsidisation occurs and thus that coal saved by the CEGB would not be exported. Conversely, we also assume that the NCB is able to supply domestic demand for coal throughout the life of any new power station, and hence the marginal source of coal supply displaced by Sizewell 'B' would be domestic rather than imported.

These two assumptions about the coal industry mean that a change in the CEGB's demand for coal would lead to a change in NCB production and employment. From the macroeconomic point of view these assumptions are important. If the Sizewell PWR displaced imports of coal, or if it released domestic coal for export, the balance of payments would improve, allowing a higher level of income and employment in the economy as a whole.

Financing
The CEGB intends to finance Sizewell 'B' out of funds accruing from depreciation charges on existing plant and from its reserves. However, this does not mean that if Sizewell 'B' is deferred the money would instead add to the CEGB's reserves. In nationalised industries the level of prices, retained profits and investment are determined by a political bargaining process between government, consumers and the industry concerned. Given this structure, it is extremely unlikely that the CEGB would be allowed to accumulate greater and greater reserves or, in other words, to earn profits which were neither passed back to the government nor on to the consumer in the form of lower prices. In practice it is likely that all the funds released by deferring Sizewell 'B' would be disbursed through these two routes, and given the tight controls on the Public Sector Borrowing Requirement (PSBR) the government would in fact probably seek to recoup all the funds not spent on Sizewell 'B'. This would be achieved by lowering the annual External Financing Limit set for the electricity supply industry so that less was borrowed from, or more returned to, central government.

Assuming that central government followed this procedure, the net effect of deferring Sizewell 'B' would be that all the funds released would become available to the Exchequer. Conversely, the construction of Sizewell 'B' or any other power station would thus be a claim on the Exchequer even if the funds were being generated internally by the CEGB. The Exchequer would be forgoing revenue it could have obtained from the CEGB. For a given size of PSBR the government would therefore have to increase taxation or reduce other public expenditure if it decided a new power station should be built. For present purposes the precise government reaction is not important. The point is that somewhere in the economy expenditure and hence employment would be reduced as a result of the decision to build a power station.

To illustrate the magnitude of the reduction involved we assume that tax levels are adjusted to finance construction. Also, to avoid unnecessary complication, we assume that any change in households' post-tax income is fully reflected in household expenditure. Hence increases in taxes lead directly to a reduction in demand for consumer goods and services. The methods used to derive the resulting changes in expenditure are explained in the Appendix. The changes depend on how much personal spending leaks abroad through imports, or back to the government in taxes.

The discussion above dealt with the financing of the construction of a power station but a broadly similar argument applies to its operating phase. If a new power station reduced the CEGB's expenditure below what it otherwise would have been to generate the same quantity of electricity the CEGB would have more funds at its disposal to disburse as lower electricity prices or as payments to the Exchequer. We assume that the additional funds accruing to the CEGB as higher capital charges are passed on to the Exchequer, and thence to taxpayers, and the remainder to consumers through lower electricity prices. In both cases the effect would be to raise household income and expenditure. Lower operating costs for the CEGB thus lead to higher household expenditure.

Final macroeconomic effects

In any national economy there are constraints which limit the level of income and employment at any given time. The extent to which an individual project, such as investment in a new power station, leads to a higher level of national income and employment depends on the extent to which it relaxes these constraints.

Our view is that the long-term constraint on the overall level of economic activity in the UK is the balance of payments. Expansion in expenditure by either the public or private sectors can only be sustained if the resulting increase in imports can be financed by an increase in exports or by foreign investments in the UK. Failure to finance substantial additional imports in these ways leads to a fall in the foreign exchange value of sterling, and in practice governments intervene to prevent such a collapse by reducing domestic spending to a level where the threat no longer exists.

This view of how the British economy operates is based on the way in which the economy has been managed during almost the entire post-war period. It must be noted that at the present time the balance of payments constraint is less important than previously, since the government has imposed a prior constraint, the size of the PSBR, on the level of domestic expenditure. However, we take the view that this situation is a temporary one and that, in view of the anticipated deterioration in Britain's balance of trade,[2] the balance of payments will re-assert itself as the main constraint on the level of economic activity. What this means is that the construction of a power station, such as Sizewell 'B', will lead to a higher level of income and employment in the economy as a whole only if it either increases exports or reduces imports, and thus relaxes the balance of payments constraint.

So far we have discussed the direct effects of the decision to build a power station and the financial implications for government, taxpayers and consumers. These affect the balance of payments in three ways:

(i)   through CEGB purchases from abroad;
(ii) through the net purchases of the CEGB's suppliers (directly and indirectly) from abroad;
(iii) through the change in the level of imports resulting from a change in household expenditure.

In addition there are two further routes through which a new power station could affect the balance of payments:

(iv) through the impact of electricity prices on producers' costs, competitiveness, and thus on imports and exports;
(v)   through export 'spin-off'.

Electricity prices are affected by a power station during both the construction (when interest on capital expenditure is financed out of current revenue) and the operational life of the station. The methods used to estimate the impact of electricity prices on foreign trade are described in the Appendix. The important point is that the effects are small, at least by comparison with the other changes in imports resulting from building a power station. This is because electricity charges amount to only about 1.0 per cent of total industrial costs on average, and the impact of Sizewell 'B' on electricity prices (on CEGB figures) is unlikely to be more than 1.0 per cent, so the overall effect on industrial costs will be of the order of magnitude of 0.01 per cent. The trading performance of some individual industries where electricity accounts for a larger part of total costs (e.g., aluminium smelting) is, of course, more sensitive to electricity prices.

Export 'spin-off' could arise from the construction of Sizewell 'B' because as the first PWR to be built in Britain it might allow British companies to gain sufficient experience to enable them to manufacture and build PWRs for other countries. However, Sizewell 'B' by itself is unlikely to generate PWR orders for British companies, particularly as most of the main components of the nuclear steam supply system will be built abroad. And even if a programme of PWRs followed Sizewell 'B', export orders are again unlikely on anything but a modest scale since, by virtue of their vastly greater experience, American, French and German manufacturers have a marked competitive advantage in the world market. Moreover, any export 'spin-off' in the nuclear industry arising from a PWR programme could easily be offset by the possible loss of exports from firms making mining machinery that might arise if the UK coal industry were to be run-down and their sound base in the British market eroded. Exports of coal-mining equipment were worth £132 million in 1981. For these reasons we regard any net export 'spin-off' as highly uncertain, and no allowance for this is included.

The balance of payments effect of a new power station is therefore defined as the sum of the changes in trade resulting from purchases by the CEGB and its suppliers, changes in household income, and changes in electricity prices.

If the net effect is a deterioration in the balance of trade the government could allow the exchange rate to depreciate sufficiently to compensate for the deterioration. Income and employment would tend to recover in the short term owing to the increase in competitiveness resulting from a lower exchange rate, but import prices would rise and the higher level of inflation would encourage higher wage increases. Higher wages lead to a loss of competitiveness which offsets the initial gain due to the devaluation. Thus, in our view, in the long term devaluation increases prices but does not improve the balance of payments.

The main alternative to devaluation is for the government to use fiscal policy to cut total expenditure in the economy. This lowers national income but reduces imports and thereby improves the balance of payments. This has been the principle method by which recent governments have managed the balance of payments, and given the limited effectiveness of devaluation is likely to remain so. We therefore define the 'final' effects of Sizewell 'B' and alternative power stations as the effect on national income and employment after the government has undertaken the necessary fiscal adjustment to restore the balance of payments to what it otherwise would have been. The methods by which these calculations are made are explained in the Appendix.

Chapter 4

The Construction Phase

The economic impact of a decision to build a new power station differs between the construction phase and the operating phase. Clarity is thus best served by considering each phase separately. In this section we compare the construction of the proposed Sizewell PWR with a coal-fired station. To ease comparability the figures presented are for stations commissioned in 1991, although in practice it is unlikely that the CEGB would commission another coal station before 1997 even if there was a moratorium on nuclear power station building.

Capital costs

Table 4.1 shows the estimated capital cost of Sizewell 'B' and a new coal-fired alternative. The figures for a coal station are based on a 1875 MWso station, but to facilitate comparison they have been scaled down to provide estimates for a station of 1110 MWso, the same as Sizewell 'B'. The estimates for a coal station and one set of estimates for Sizewell 'B' are those presented by the CEGB, including the start-to-finish allowance and the construction period extension.[1] The second set of estimates for Sizewell 'B' assumes that the differential between PWR and coal station capital costs will be larger, in line with American experience (see Chapter 2). The additional costs have been allocated proportionately to the items most likely to be affected by over-runs in the design and construction of the nuclear island.[2]

Direct effects

The Chairman of the CEGB has stated that the proportion of Sizewell 'B's total capital cost going to imports is likely to be 6–10 per cent.[3] Most of the nuclear steam supply system is likely to be imported, since there is no UK manufacturing capacity at present. We assume that 10 per cent of the capital expenditure on Sizewell 'B' goes directly on imports and that 58 per cent of these will be related to the nuclear steam supply system and 7 per cent to other electrical plant. The remaining imports are assumed to be payments to the Westinghouse and Bechtel Corporations in the USA for the purchase of the PWR design and for services provided to help implement the design in the UK. In addition, we assume that 74 per cent of the cost of the initial nuclear fuel for the reactor will consist of imports.[4] By contrast, none of the major contracts for the construction of a coal-fired station is likely to be placed abroad.

These assumptions about imports enable the capital expenditure to be allocated between the main industries concerned, as shown in Table 4.2. One feature worth noting is that despite being substantially cheaper overall,

Table 4·1 Estimated capital cost of new power stations (excluding interest during construction)

£m, March 1982 prices

	Sizewell 'B'		Coal [a]
	CEGB costs	Alternative costs	
Civil engineering and building	257[b]	424	152
Boiler/nuclear steam supply system	161	252	255
Turbine generators	146	146	142
Other mechanical plant	95	95	69
Other electrical plant	108	112	74
NNC costs	332	539	—
CEGB engineering	73[c]	120	45
Sub-total	1172	1688	737
Initial fuel	76	76	—
TOTAL	1248	1764	737

[a] Notional costs for a station with the same output as Sizewell 'B'.
[b] Includes visitor centre (£3m).
[c] Includes two PWR simulators (£22m).

Source: Adapted from CEGB, *Proof of Evidence, P8* (1982), A. Wilson, *On: Construction Time, Cost and Operating Performance of PWR, AGR and Coal-fired Generating Plant*, Vol. 2, Table 21, p. 117.

a coal station would provide nearly twice as much work in mechanical engineering, mainly because the boilers for a coal station would be produced in Britain. The higher expenditure on a PWR as compared with a coal station mainly accrues to the construction industry and to the engineering design centres of the NNC and the CEGB.

In the first part of Table 4.3, expenditure by UK contractors to the CEGB has been allocated between two headings: value-added by the contractors and their UK suppliers, and purchases from abroad. The calculations have been based upon the most recent input-output tables for the UK economy, and enable the estimation of the total domestic value-added content of Sizewell 'B' and a coal-fired alternative, and of total imports including imports by UK contractors and their suppliers.[5] Taking into account these imports, the overall import bill associated with Sizewell 'B' is likely to be at least £370 million and possibly as much as £483 million. Imports associated with the construction of a coal-fired station are lower, at £173 million.

Table 4·2 Estimated distribution by industrial sector of capital expenditure on new power stations

£m, March 1982 prices

	Sizewell 'B'		Coal
	CEGB costs	Alternative costs	
Mechanical engineering [a]	188	249	324
Electrical engineering [b]	246	246	216
Construction	257	424	152
NNC and CEGB engineering	364	600	45
Nuclear fuel production	20	20	—
Imports [c]	173	225	—
TOTAL	1248	1764	737

[a] Includes boiler/nuclear steam supply system and other mechanical plant (excluding imports).
[b] Includes turbine generators and other electrical plant (excluding imports).
[c] Assumed to be 10% of capital cost of Sizewell 'B', plus the import component of the initial fuel.

Source: Adapted from CEGB, *Proof of Evidence, P8*.

The direct impact on employment in the UK, shown in the second part of Table 4.3, is estimated on the basis of value-added per employee in the UK by the contractors, their suppliers and the CEGB itself. Sizewell 'B' will support about 8,400 jobs in an average year for seven-and-a-half years. If the alternative costings proved more accurate then the number would rise to over 10,000 a year for up to nine years. A coal-fired station would support roughly 6,500 jobs a year for six-and-a-half years. The greater estimated

Notes to Table 4.3
[a] The higher average annual employment figure for Sizewell 'B' using the alternative costs does not necessarily imply higher peak employment.
[b] We have assumed that NNC and CEGB employment is related to expenditure on a new power station in the same manner as other sectors. If these jobs, mostly in design facilities, were in fact unrelated to the construction of specific stations this would make a substantial difference to the relative employment-supporting potential of nuclear and coal-fired units. For example, if it were assumed that none of the NNC and CEGB jobs was related to specific power station projects then Sizewell 'B' and a coal-fired station on CEGB costs would support the same average annual number of direct jobs.

Note: See Appendix for details of calculations.

Table 4.3 Estimated effects on value-added and employment of power station construction, by sector

	VALUE-ADDED IN UK (£m, March 1982 prices)			AVERAGE ANNUAL EMPLOYMENT		
	Sizewell 'B'		Coal	Sizewell 'B'		Coal
	CEGB costs	Alternative costs		CEGB costs	Alternative costs [a]	
Mechanical engineering	65	87	113	800	900	1500
Electrical engineering	98	98	86	1400	1200	1300
Construction	143	235	84	1400	1900	900
NNC and CEGB engineering [b]	288	474	36	2200	3100	300
Nuclear fuel production	10	10	—	100	100	—
Other industries	251	344	225	2500	2900	2500
TOTAL	854	1247	544	8400	10100	6500
plus Imports	370	483	173			
plus Indirect taxes	24	34	20			
equals Total expenditure	1248	1764	737			

employment provided by the construction of Sizewell 'B' is a result of the much greater capital expenditure associated with a nuclear station. It must be stressed that these calculations are illustrative only, in that they are based on industry averages and projected increases in labour productivity and import penetration. In practice, individual factories would differ from the average experience of their industry and hours worked might increase, rather than the number of jobs.

The regional impact of the construction of Sizewell 'B' or a coal-fired alternative is impossible to gauge accurately without detailed knowledge of which companies would receive contracts. However, several of the factories potentially able to supply boilers, turbine generators and other major mechanical and electrical equipment are in areas with above-average unemployment. With a PWR many of the jobs would be at the CEGB engineering centre at Barnwood in Gloucestershire and at the National Nuclear Corporation which employs staff in Leicestershire and Lancashire. The construction jobs associated with Sizewell 'B' would be in Suffolk, an area with below-average unemployment, whereas a coal-fired station would be likely to be located on or near a coalfield where unemployment is mostly above average. On balance, a coal-fired station would probably do more to narrow regional differences in employment opportunities, whereas Sizewell 'B' might well benefit areas of low unemployment.

The employment figures represent an average over the construction

Table 4·4 The projected annual distribution of capital expenditure

(per cent)

	Sizewell 'B'	Coal
Pre-construction period	13	8
1984/85	10	11
1985/86	14	16
1986/87	21	23
1987/88	16	20
1988/89	12	12
1989/90	9	4
1990/91	4	2
1991/92	1	2
Post-construction period	0	2

Source: Based on a communication from Mr P. E. Watts, CEGB Finance Department,
          9 December 1982.

period, although in fact expenditure on power station construction rises to a peak about three years after work has started, as Table 4.4 demonstrates. We have not calculated the changes in employment on a year-by-year basis because the incidence of imports during construction is not known, but it is likely that the direct employment effects would be spread over the construction period in a similar manner, with a concentration in the middle years.

Final effects
The burden of financing the construction of a power station falls ultimately on the Exchequer (see Chapter 3). Since we assume this requires higher levels of taxation, expenditure and thus employment will be lower elsewhere in the economy. In effect, there is a shift of expenditure from consumption to investment.

Some of this reduction in consumers' expenditure would be absorbed by a reduction in imports. This is important because a high proportion of consumers' expenditure goes on imports, and given the average level of import penetration likely to prevail during the construction period[6] the reduction would probably be larger than the increase in imports resulting directly from power station construction.[7] In other words, the construction of power stations would be likely to improve the balance of trade because of the relatively low import content of power station construction compared to consumer spending.[8] As explained in Chapter 3, relaxation of the balance of payments constraint would enable the government to expand consumer spending through tax cuts up to the point where the balance of payments was restored to the level it would have been in the absence of power station construction. The final effects on national income and employment are shown in Table 4.5.

If Sizewell 'B' were built within the CEGB's cost estimates it would lead to the smallest increase in national income. This is because it has the least beneficial impact on the balance of payments, given that large parts of the nuclear steam supply system would be imported. If there were large cost over-runs with Sizewell 'B', as implied by the alternative estimates, the increase in national income during the construction period would be the greatest. This is because a high-cost PWR would divert the largest amount of income away from import-intensive consumer spending, and also because we assume major cost over-runs would be reflected mainly in UK construction, design and engineering costs rather than the import content of the project. Despite the relatively small amount of capital investment involved, a coal-fired station would also lead to a substantial increase in national income, at least in part because few major components would be imported.

So far as employment is concerned, the construction of Sizewell 'B' within the CEGB's cost estimates would have the smallest net effect on the economy as a whole, an increase of about 1,100 jobs. If there were large cost over-runs with the PWR there would be a net increase of about 1,900 jobs in the economy. A coal-fired station would generate the largest net increase in employment, at about 2,600 jobs in an average year during the construction phase. This is shown in the bottom part of Table 4.5. These figures can only be illustrative, of course, because they are based on anticipated average earnings and productivity and import penetration in the

Table 4·5   Estimated final effects of power station construction on the UK economy

	Sizewell 'B'		Coal
	CEGB costs	Alternative costs	
NATIONAL INCOME (£m, March 1982 prices)			
Income directly or indirectly from power station construction	+ 854	+1247	+ 544
Income in rest of economy	− 748	− 986	− 361
Net Change	+ 106	+ 261	+ 183
EMPLOYMENT (average annual)			
Construction period (years)	(7.5)	(8.75)	(6.7)
Directly and indirectly in power station construction	+8400	+10100	+6500
Rest of economy	−7300	−8200	−3900
Net Change	+1100	+1900	+2600

Note: Methods of calculation are explained in the appendix.

industries concerned, rather than precise data for the actual factories and offices involved.

It must be emphasised that these increases in income and employment would be achieved only at the expense of a very large diversion of expenditure from household consumption into investment in power stations. This point is illustrated also by Table 4.5, which shows that the reduction in income and employment in the rest of the economy, outside power station construction, would be substantially smaller for a coal-fired station than for Sizewell 'B'. The reasons for this are again the low import content of the coal station and the smaller increase in taxation necessary to finance its construction. Whether or not the greater diversion of expenditure from consumption to investment in the case of a PWR is desirable depends on the value placed on the consequences of nuclear power station construction (which are supposed to be cheaper electricity in later years) relative to higher consumer spending in the immediate future.

Summary
If the CEGB were to begin constructing Sizewell 'B' for commissioning in 1992 and if it were built within its budgeted cost, national income and

employment would be lower during the construction period than if a coal-fired station were built instead. Moreover, the construction of a coal-fired station would require a smaller reduction in consumption to finance construction. It also seems likely that a coal-fired station would do most to diminish regional disparities in unemployment.

Chapter 5

The Operating Phase

During the 35 or 40 years in which a new power station would be in operation[1] there would be economic consequences for electricity consumers, taxpayers and those directly or indirectly deriving income from its operation. The CEGB focuses on the consequences for electricity consumers of the costs of operating the various options, and ignores the other consequences. Our approach is different in that we also consider the effects on the rest of the economy.

The CEGB assessment discounts estimated operating costs back to the year of commissioning, thus arriving at one figure for each option which sums up its relative merits. An advantage of this approach is that all future years are considered, with more importance being given to the immediate future than to later years. The weight attached to earlier and later years is rather arbitrary, however, and we believe greater clarity is achieved using undiscounted costs. We therefore distinguish between three distinct phases in the operating life of Sizewell 'B' and present figures for a representative year in each case. It is really a political decision as to how much importance should be attached to the costs and benefits in each of these periods, and we feel it is advantageous to show explicity how costs and benefits alter over the power station's lifetime. The three periods covering different phases of operation are as follows:

(i) *c. 1992–1996*

During these years new generating plant is not needed on capacity grounds, according to the CEGB central estimates. A new station would replace existing serviceable coal and oil stations, and hence we compare the consequences of operating Sizewell 'B' not only with a new coal-fired station but also with existing stations. The representative year is taken as 1996, by which time new stations would have reached 'settled-down' availability.

(ii) *c. 1997–2010*

By 1997, the CEGB estimates that new plant will be needed on capacity grounds. We therefore compare Sizewell 'B' with a new coal-fired station. During this period both stations would be used to supply base-load power or, in other words, they would be used to the limits of their availability. The representative year is taken to be 2003.

(iii) *c. 2011–2030*

In this third period a new coal-fired station would be likely to operate at below its maximum load factor because more efficient stations would be in use. If a large number of new nuclear stations were built, Sizewell 'B' might also no longer be used to supply base load. However, we assume that Sizewell 'B' continues to operate to the limits of its availability during this third period. The representative year is taken to be 2020.

The CEGB does not present a breakdown of operating costs for each of these years so estimates have been derived from its *Proofs of Evidence*. One set of figures is based entirely on CEGB central estimates of costs and a second set based on alternative assumptions is also presented. These alternative assumptions were discussed in Chapter 2. They are:–

(i)   A 44 per cent higher capital cost for Sizewell 'B', in line with American experience.
(ii) A lower load factor for Sizewell 'B' (58 per cent instead of 64 per cent), again in line with American PWR experience.
(iii) An increase in the real price of coal at only half the rate assumed by the CEGB.

As in the construction period, Sizewell 'B' is compared with a notional coal station with the same output (1110 MWso).

A new power station as a replacement for older stations
On CEGB estimates, no new power station is needed on capacity grounds until 1997. If Sizewell 'B' or a new coal-fired alternative is commissioned before then, older power stations will be retired prematurely. Some of these older stations burn coal, others oil. The operating costs in 1996 of Sizewell 'B', a new coal station and the older stations that would be replaced are compared in Table 5.1. A point to note is that the accounting costs are of two kinds.

First, there are the operating costs consisting of wages, fuel and other materials, etc. These are a direct call on the economy's productive capacity and foreign exchange earnings. Second, the cost to the consumer also includes a large capital charge, nominally to repay the previously-incurred construction cost of the power station. Capital charges accrue to the Exchequer, either directly as repayments from the CEGB or indirectly if they are used to pay for new investment which the Exchequer would otherwise have had to finance. The Exchequer is able to use this revenue to reduce taxes, and to increase household income. In essence, capital charges thus transfer income from electricity consumers to taxpayers, most of whom are the same people. Therefore, although the accounting cost is a true measure of the overall cost of a power station's electricity, once the station is built it is the operating costs which affect the economy most directly.

As Table 5.1 shows, the main cost of Sizewell 'B' in the mid 1990s would be capital charges, whereas the main cost of a new coal station and older stations would be fuel. On balance, CEGB figures indicate that, in terms of accounting costs, Sizewell 'B' would be around £100 million a year cheaper to run than older stations, and around £60 million cheaper than a new coal station. If the alternative costs proved more accurate, Sizewell 'B' would be £55 million cheaper than older stations, but would actually be more expensive than a new coal station.

Table 5·1 Estimated annual operating costs in 1996: Sizewell 'B' and alternative stations

(£m, March 1982 prices)

	CEGB costs			Alternative costs		
	Sizewell 'B'	New coal	Old stations[a]	Sizewell 'B'	New coal	Old stations[a]
Fuel used: nuclear[b]	24	—	—	21	—	—
coal[c]	—	151	86	—	131	68
oil[d]	—	—	129	—	—	117
Other operating costs (net)[e]	6	8	18	6	8	18
Sub-total	30	159	233	27	139	203
Fuel saved[f]: coal	0	13	0	0	19	0
oil	0	17	0	0	28	0
NET OPERATING COST	30	129	233	27	92	203
Capital charges[g]	131	93	37	170	93	37
ACCOUNTING COST	161	222	270	197	185	240

Because of its lower operating costs Sizewell 'B' would support fewer jobs in UK fuel industries and their suppliers than either a new coal station or older stations. Table 5.2 illustrates this point. In particular, Sizewell 'B' would reduce production and employment in the coal industry and its suppliers, and as British Rail would lose some of its coal-carrying business there would be job losses on the railways too. Using estimated relationships between changes in output and employment, it appears that running Sizewell 'B' instead of older stations would reduce employment in the power industry and its suppliers by 6–7,000 in total. Compared with a new coal-fired station, Sizewell 'B' would reduce total employment in these industries by 9–10,000.

However, offsetting these job losses, Sizewell 'B' would at the same time allow an increase in income in the rest of the economy. Part of this would arise because the additional capital charges associated with a PWR could be returned to taxpayers *via* the Exchequer, and on the basis of CEGB costings an increase in households' real income would result from the lower electricity prices made possible by Sizewell 'B's lower operating costs. A further small increase in income from higher exports and lower imports would be likely to result from the lower cost of electricity to industry.

The main way in which Sizewell 'B' would raise income and employment in the economy as a whole would be by reducing the CEGB's use of oil in older power stations. This oil could instead be exported, thereby improving the balance of payments and allowing fiscal expansion by the government to increase national income and expenditure and restore the trade balance to what it otherwise would have been. However, a new coal-fired station would displace the same oil stations, so the same benefit to the economy would also accrue from a new coal station.

The final changes in national income and employment associated with Sizewell 'B', including the effect of the macroeconomic adjustment in response to the balance of payments, are shown in Table 5.3. Two sets of figures are presented: one comparing Sizewell 'B' with older stations and the other comparing it with a new coal station. The table demonstrates that the early completion of Sizewell 'B' as a replacement for older stations would increase income and employment in the economy as a whole. The alterna-

---

Notes for Table 5.1

[a] Providing the same electricity output as Sizewell 'B'.
[b] Nuclear fuel cycle costs are taken from CEGB, *P9*, Tables 4–7.
[c] CEGB costs: £53 per tonne (source: CEGB, *P6*, Table 22, plus an addition of 11% to pithead prices to allow for transport costs).
     Alternative costs: £46 per tonne. Quantities: communication from CEGB, 4 July 1983.
[d] £117 per tonne (source: CEGB, *P6*, Table 24). Quantities: communication from the CEGB, 4 July 1983.
[e] Fixed and variable operating costs less variable operating savings in other stations (sources: CEGB, *P4*, Appendix 4, pp. 141–2 and p. 155).
     The average net variable cost savings given on p. 155 are assumed to decline linearly from the first year of operation to zero in the final year.
[f] Savings arising from the higher load factor of coal stations.
[g] Capital charges for Sizewell 'B' and new coal station include provision for annual connection and reinforcement charges, and Sizewell 'B' includes provision for decommissioning costs (source: CEGB, *P4*, Appendix 4, pp. 140–1). Capital charges of older stations retired before the end of their design lives are included with Sizewell 'B' and new coal. Capital charges for older stations are estimated at two-thirds of replacement costs (for a new coal station) to allow for the lower thermal efficiency of older stations. On an annuitised basis these costs are estimated at £37 million per annum.

Table 5·2 Estimated effect in 1996 of operating Sizewell 'B' on value-added and employment in UK fuel industries and their suppliers

	CEGB costs		Alternative costs	
	Value-added (£m, March 1982 prices)	Employment	Value-added (£m, March 1982 prices)	Employment
SIZEWELL 'B' INSTEAD OF OLDER STATIONS				
Coal industry	− 36	− 4800	− 38	−3900
Railways	− 2	− 100	− 2	− 100
Nuclear fuel production	+ 5	+ 300	+ 4	+ 100
Oil industry	0	0	0	0
Electricity industry	− 12	− 700	− 12	− 700
Other industries	− 21	− 1400	− 22	−1500
TOTAL	− 66	− 6700	− 70	−6100
SIZEWELL 'B' INSTEAD OF A NEW COAL STATION				
Coal industry	− 58	− 7700	− 63	−6400
Railways	− 3	− 200	− 3	− 200
Nuclear fuel production	+ 5	+ 300	+ 4	+ 200
Oil industry	0	0	0	0
Electricity industry	− 2	− 100	− 2	− 100
Other industries	− 36	− 2400	− 38	−2600
TOTAL	− 94	−10100	−102	−9100

Note: See the Appendix for details of the calculations.

tive cost estimates for Sizewell 'B' – based on higher capital costs, lower availability and lower coal prices – reduce the potential gain, though it remains large even on these assumptions. Coupled to the rise in national income and employment there would be a substantial shift from fuel industries (especially coal mining) to the rest of the economy.

Nevertheless, Table 5.3 also demonstrates that during the mid 1990s

Table 5·3 Estimated final effect on the UK economy in 1996 of operating Sizewell 'B'

	INSTEAD OF OLD STATIONS		INSTEAD OF NEW COAL STATION	
	CEGB costs	Alternative costs	CEGB costs	Alternative costs
NATIONAL INCOME (£m, March 1982 prices)				
Income in the electricity and fuel industries and their suppliers	− 66	− 70	− 94	− 102
Income in rest of economy	+ 246	+ 208	+ 20	− 25
Net change	+ 180	+ 138	− 74	− 127
EMPLOYMENT				
Electricity and fuel industries and their suppliers	− 6700	− 6100	−10100	− 9100
Rest of economy	+15300	+13800	+ 1300	− 1700
Net change	+ 9600	+ 7700	− 8800	−10800

Note: See the Appendix for the methods of calculation.

Sizewell 'B' would have no advantage over a new coal-fired alternative. CEGB costs indicate that Sizewell 'B' would lead to a lower level of national income than a new coal station, and using the alternative cost assumptions the loss is greater. Moreover, Sizewell 'B' would support substantially fewer jobs than a new coal station. This is because a new coal station would, like Sizewell 'B', release oil for export and thus allow an expansion in the economy, but it would also avoid the major job losses in fuel industries that would be associated with a PWR.

Furthermore, the completion of Sizewell 'B' ahead of demand would undoubtedly widen regional unemployment differentials. Roughly three-quarters of the jobs lost by operating Sizewell 'B' would be in regions of above average unemployment.[2] Half would be in areas of very high unemployment. In contrast, offsetting gains could be expected to be fairly evenly spread throughout the country.

The period of base-load operation
Until 1996 the CEGB case for Sizewell 'B' depends on it being cheaper to run than existing fossil-fuel stations. After 1996, if the CEGB central forecasts are correct, it will be necessary to operate some new stations to provide adequate capacity. The CEGB case for Sizewell 'B' after 1996 is that it would be cheaper to run than other plausible alternatives, notably a new coal station. We compare Sizewell 'B' with a new coal-fired station of the same age. During the period up to roughly 2010 both stations would be operated at the limits of their availability since they would be relatively new and more efficient than older stations.

Table 5.4 presents a breakdown of the operating costs of the two stations during this period of base-load operation. The savings of coal and oil at other power stations resulting from the higher load factor expected to be achieved by a coal station (72 per cent compared to 64 or 58 per cent) are included in these accounts. The main cost of Sizewell 'B' would still be capital charges, and the main cost of a coal station would still be fuel. On CEGB assumptions about costs, including coal prices, Sizewell 'B' would be cheaper in accounting terms. The situation is reversed by the alternative assumptions, highlighting the extent to which the CEGB's economic case is sensitive to assumptions about construction costs, performance and coal prices.

Sizewell 'B', like all nuclear power stations, has operating costs which are substantially below those of a coal-fired alternative. Hence relatively little is spent on UK fuel industries and their suppliers. Table 5.5 traces the consequences for these industries. Running a PWR rather than a new coal station would greatly reduce employment in the coal industry, which would lose as many as 6,500 jobs, and there would be further job losses in other industries, especially those supplying the coal industry.

Imports associated with Sizewell 'B' include uranium ore. Imports associated with a coal station are mainly the imports by the coal industry's suppliers, though these are partly offset by the exports of the additional oil saved. On CEGB costs, Sizewell 'B' would lead to a net reduction in imports by the CEGB and its suppliers, compared to a new coal station, and the slightly lower electricity prices associated with Sizewell 'B' would also improve the balance of payments by reducing industrial costs. However,

Table 5·4 Estimated annual operating costs 1997–2010: Sizewell 'B' and a coal-fired alternative

(£m, March 1982 prices)

	CEGB costs		Alternative costs	
	Sizewell 'B'	Coal	Sizewell 'B'	Coal
Fuel used [a]	33	170	30	139
Other operating costs (net) [b]	9	10	9	10
Sub-total	42	180	39	149
Fuel saved: [c] coal	0	21	0	27
oil [d]	0	9	0	16
NET OPERATING COST	42	150	39	106
Capital charges [e]	94	56	133	56
ACCOUNTING COST	136	206	172	162

[a] Nuclear fuel: see CEGB, *P9*, Tables 4–7; coal = £60 per tonne (CEGB costs), £49 per tonne (alternative costs). See Table 5·1, notes b and c.
[b] See Table 5·1 note e.
[c] Savings arising from the higher load factor of coal stations.
[d] £146 per tonne. See Table 5·1, note d.
[e] Including provisions for annual connection and reinforcement charges, and decommissioning of Sizewell 'B' (source: CEGB, *P4*, Appendix 4, pp. 140–141).

Note: The representative year is taken to be 2003.

additional consumer spending induced by lower electricity prices and lower taxes (financed by Sizewell 'B's higher capital charges) is likely to cause a larger offsetting deterioration in the trade balance because of the high import content of consumer expenditure.

The final effects on the UK economy, shown in Table 5.6, are calculated in the same way as previously, taking account of the macroeconomic adjustment necessary in response to changes in the balance of payments. On CEGB costs Sizewell 'B' will tighten the balance of payments constraint so that overall national income will be lower than if a coal station were built. Using the alternative cost assumptions national income would be lower still.

Associated with the lower national income arising from Sizewell 'B' would be a large net decline in employment, shown in the second part of

Table 5·5 Estimated effect of operating Sizewell 'B' instead of a new coal-fired station on value-added and employment in UK fuel industries and their suppliers, 1997–2010

	CEGB costs		Alternative costs	
	Value-added (£m, March 1982 prices)	Employment	Value-added (£m, March 1982 prices)	Employment
Coal industry	−60	−6400	−61	−5000
Railways	− 4	− 300	− 3	− 200
Nuclear fuel production	+ 6	+ 300	+ 6	+ 300
Oil industry	0	0	0	0
Electricity industry	− 1	− 100	− 1	− 100
Other industries	−37	−2400	−35	−2200
TOTAL	−96	−8900	−94	−7200

Notes: See the Appendix for details of the calculations.
        The representative year is taken to be 2003.

Table 5.6. The problem is that Sizewell 'B' would require substantially fewer workers (in the CEGB and suppliers such as the coal industry) than a coal station in order to generate the same amount of electricity. The resulting savings, passed on to households (as lower electricity prices or tax cuts financed out of additional Exchequer revenue from the CEGB), mean that households' real income would be higher. However, after allowing for adjustments in fiscal policy to restore the balance of payments to what it otherwise would have been, the additional household expenditure would generate insufficient jobs in the UK to offset the losses in the fuel and associated industries. In other words, Sizewell 'B' would lead to an increase in real income for those in work, but only at the expense of greater unemployment, mostly among miners. On balance, we estimate that during the period of base-load operation, employment would be 7–8,000 lower with Sizewell 'B' instead of a new coal station.

The later years of operation
After about 2010 a coal-fired station commissioned in 1997 or earlier would probably cease to supply base-load power and begin to operate at below its maximum potential availability. This would occur because sufficient newer and presumably more efficient power stations would have been constructed to supply all base-load needs, assuming the CEGB central estimate of demand proves correct. Whether the Sizewell PWR would also cease to be a

base-load station would depend on how much new nuclear capacity was installed before 2010. If a large number of nuclear stations were built after Sizewell 'B', these newer (and again more efficient) stations would supply base load.

Table 5·6 Estimated final effect on the UK economy of operating Sizewell 'B' instead of a new coal-fired station, 1997–2010

	CEGB costs	Alternative costs
NATIONAL INCOME (£m per year, March 1982 prices)		
Income in the electricity and fuel industries and their suppliers	− 96	− 94
Income in rest of economy	+ 24	− 6
Net change	− 72	− 100
EMPLOYMENT		
Electricity and fuel industries and their suppliers	−8900	−7200
Rest of economy	+1500	− 400
Net change	−7400	−7600

Notes: See the Appendix for the methods of calculation.
The representative year is taken to be 2003.

In order to keep the decision about Sizewell 'B' separate from a larger PWR programme, our estimates of the operating economics after 2010 assume that no further nuclear stations are built after Sizewell 'B'. In these circumstances Sizewell 'B' would continue as a base-load station throughout its life. This differs from the CEGB assumption of a 'high nuclear background' (i.e. a major programme of nuclear construction) which is used in their main evaluation of Sizewell 'B', and it should be noted that our assumption is more favourable to the economics of Sizewell 'B' than to a coal-fired alternative. In the comparison which follows, the representative year is taken to be 2020, when it is assumed that a coal-fired station would be operated at a 10.5 per cent load factor[3] while Sizewell 'B' continued to operate at maximum availability.

Table 5.7 presents the estimated annual operating costs of Sizewell 'B' and a coal-fired station of the same age. Because we assume Sizewell 'B'

Table 5·7 Estimated annual operating costs, 2011–2030: Sizewell 'B' and a coal-fired alternative

(£m, March 1982 prices)

	CEGB costs		Alternative costs	
	Sizewell 'B'	Coal	Sizewell 'B'	Coal
Fuel[a]	45	33	41	23
Other operating costs (net)[b]	16	15	16	15
Sub-total	61	48	57	38
Fuel used in other stations[c]	0	182	0	113
NET OPERATING COSTS	61	230	57	151
Capital charges[d]	94	56	133	56
ACCOUNTING COST	155	286	190	207

[a] Nuclear fuel: see CEGB, *P9*, Tables 4–7; Coal = £81 per tonne (CEGB costs). £57 per tonne (alternative costs). See Table 5·1, notes b and c. The coal-fired station is assumed to have a load factor of 10.5%. Transport and handling costs of coal are assumed to be 11% of pithead price.
[b] See Table 5·1 note e.
[c] Cost of coal burned in newer stations.
[d] See Table 5·1 note e.

Note: The representative year is taken to be 2020.

would be operated at a higher load factor the costs include, in the case of the coal station, the cost of the coal burned at newer power stations in order to generate the same quantity of electricity as Sizewell 'B'. Taking the CEGB-based cost estimates first, Sizewell 'B' appears to be the substantially cheaper option. This is mainly because of the large increase in the real price of coal which the CEGB anticipates. The cost advantage of Sizewell 'B' largely disappears, however, if the alternative cost and performance assumptions are used, though the alternative assumptions are not sufficient to tip the balance in favour of a coal station, as was the case in the preceding period. Nevertheless, given that the CEGB believes that the gains from a PWR should be very large during this final period, it is an important conclusion that alternative assumptions indicate that such a gain might never materialise.

The estimated direct impact of Sizewell 'B' compared to a coal-fired alternative is shown in Table 5.8. As in the period of base-load operation, the main impact of Sizewell 'B' is on the coal industry and its suppliers. Coal mining could expect to lose between 3,200 and 4,900 jobs if Sizewell 'B' were built rather than a coal station. Firms supplying the coal industry would also expect to lose substantial numbers of jobs. Sizewell 'B' would have a favourable direct effect on the balance of trade, however, because imports of uranium ore would be more than offset by the reduction in imports by the coal industry and its suppliers.

Table 5·8 Estimated effect of operating Sizewell 'B' instead of a new coal-fired station on value-added and employment in UK fuel industries and their suppliers, 2011–2030

	CEGB costs		Alternative costs	
	Value-added (£m, March 1982 prices)	Employment	Value-added (£m, March 1982 prices)	Employment
Coal industry	− 62	−4900	− 65	−3200
Railways	− 5	− 300	− 3	− 200
Nuclear fuel production	+ 7	+ 300	+ 6	+ 200
Electricity industry	− 1	0	+ 1	0
Other industries	− 54	−2900	− 39	−2100
TOTAL	−113	−7800	−100	−5300

Notes: See the Appendix for details of the calculations.
The representative year is taken to be 2020.

The estimate of the final impact of Sizewell 'B' during this last period of operation is shown in Table 5.9. Sizewell 'B' would reduce the CEGB's expenditure on fuel and increase revenue in the form of capital charges so that the CEGB could either lower electricity prices or pay more to the Exchequer, allowing a reduction in taxes. The disposable income of households would therefore increase, although much of this would again be likely to be spent on imports. The favourable direct impact of Sizewell 'B' on the balance of payments would be more than offset by these consumer imports. As this would require deflation by the government to reduce imports, national income in total would be lower than for a coal station. Comparisons of the overall employment change associated with a PWR and a coal station are also unfavourable to the PWR. Sizewell 'B' would lead to a net job loss of 3–4,000 which, as in the previous period, would be overwhelmingly concentrated in the coal industry.

Table 5·9 Estimated final effect on the UK economy of operating Sizewell 'B' instead of a new coal-fired station, 2011–2030

	CEGB costs	Alternative costs
NATIONAL INCOME (£m, March 1982 prices)		
Income in the electricity and fuel industries and their suppliers	− 113	− 100
Income in rest of economy	+ 76	+ 37
Net change	− 37	− 63
EMPLOYMENT		
Electricity and fuel industries and their suppliers	−7800	−5300
Rest of economy	+4000	+1900
Net change	−3800	−3400

Notes: See the Appendix for the methods of calculation.
The representative year is taken to be 2020.

Summary

Throughout its operating life, Sizewell 'B' would lead to large job losses in the coal industry in particular, though these might be partly offset by a small increase in employment elsewhere in the economy. On balance, employment will be much lower if Sizewell 'B' is built instead of a coal station.

The consequences for national income of operating Sizewell 'B' instead of a new coal station are also unfavourable. The loss of income in the fuel and associated industries would only be partially offset by an increase in real incomes in the rest of the economy made possible by the lower operating costs of a nuclear station.

The completion of either Sizewell 'B' or a new coal station ahead of demand, leading to the early retirement of existing serviceable power stations, would however increase national income and employment. This would occur mainly because the usage of oil-fired power stations would be reduced and the oil could be exported instead, with beneficial effects on the balance of payments and the overall level of activity in the economy. A new coal station completed ahead of demand would nevertheless increase national income and employment by more than Sizewell 'B'.

Chapter 6

Conclusions and Policy Recommendations

The CEGB claims that the Sizewell 'B' nuclear power station would produce cheaper electricity than any alternative generating capacity. It also claims that since electricity from Sizewell 'B' would be substantially cheaper than that from existing stations, the construction should begin as soon as possible. However, we have argued that, whatever the merits to the CEGB of building and operating Sizewell 'B', such a large public investment should be evaluated in terms of its impact on the economy as a whole. It is by no means certain that the option which is most attractive to the CEGB will be the best when the interests of other firms and households are also taken into account.

We therefore traced the impact of the decision to build and operate Sizewell 'B' on other industries and on the economy as a whole. This exercise was undertaken separately for the construction phase and for three different periods during the operating life of the power station. Table 6.1 which compares Sizewell 'B' with a 'notional' coal-fired station with the same output, summarises the impact on national income and employment. The table also includes estimates of the effect on electricity prices.[1] As in the rest of this study, two sets of estimates are presented: one based entirely on CEGB costs, and other using alternative costs less favourable to nuclear power. It must be stressed that the estimates are based on assumptions (discussed in Chapter 3) about how the government responds to changes in its revenue and expenditure and to the balance of payments. The precise impact on the UK economy will depend on exactly how in practice the goverments responds to changes in these variables.

Table 6.1 nevertheless provides the proper basis on which a choice between a PWR and a new coal-fired station should be made. The most important point it demonstrates is that, during the construction phase and throughout its operating life, Sizewell 'B' would lead to lower levels of employment in the UK economy than a new coal-fired station. This conclusion is particularly robust since it is unaffected by the large differences in capital cost, operating performance and coal prices embodied in the two sets of costings presented. Nearly all the net decline in employment would fall on the coal industry, where employment would be up to 8,000 lower if Sizewell 'B' were built.

The operation of Sizewell 'B' would transfer spending from the CEGB to consumers, whose expenditure has a particularly high import propensity. The resulting net deterioration in the balance of payments would require the government to reduce the overall level of economic activity in order to curb imports. As Table 6.1 shows, Sizewell 'B' would therefore lead to lower levels of national income than a coal-fired alternative. The only exception,

Table 6·1 Estimated average annual effect of building and operating Sizewell 'B' instead of a new coal-fired station (summary table)

	Construction phase 1984–1991	Operating phase		
		1992–1996	1997–2010	2011–2030
CEGB COSTS				
Electricity prices (% change)	+ 0.3	− 0.7	− 0.6	− 0.8
National income (£m, March 1982 prices)	− 10	− 74	− 72	− 37
Employment	−1500	− 8800	−7400	−3800
ALTERNATIVE COSTS				
Electricity prices (% change)	+ 0.4	+ 0.1	+ 0.1	− 0.1
National income (£m, March 1982 prices)	+ 9	− 127	− 100	− 63
Employment	− 700	−10800	−7600	−3400

ironically, would be if the construction costs of Sizewell 'B' were substantially higher than planned, in which case national income might rise during the construction phase because income would be diverted from consumers and the imports they purchase towards UK industries with a lower import propensity.

Offsetting the unfavourable change in employment and national income, Sizewell 'B' offers the possibility of lower electricity prices during its operating life, assuming CEGB estimates of costs. However, this gain is distinctly uncertain and would not occur if the alternative costs proved more accurate. The benefit to industry from slightly lower electricity prices would do hardly anything to mop up the additional unemployment, because the effect on overall industrial costs and thus international competitiveness would be so small. In essence, therefore, operating Sizewell 'B' would allow slightly higher levels of real income for those in work, but only at the expense of substantially higher unemployment, mainly among miners.

Given that a shortage of employment is the main problem facing the economy, and that there is little prospect of an increase in national income from Sizewell 'B', the macroeconomic evaluation of alternative power stations does not favour the PWR.

Our first recommendation is therefore that if a new power station is built it should be a coal-fired station rather than a PWR.

A further, separate issue is the timing of the construction of a new power station. Chapter 2 showed that there is no need on capacity grounds to commission a new power station until at least 1996 and possibly even later, depending on the growth of the economy, so that the CEGB could defer ordering a new station for roughly five years. The CEGB accepts this point. Table 6.2 summarises the economic effects of a deferral, which would mean operating existing serviceable power stations for longer.

Table 6·2 Estimated effects in 1996 of deferring new power stations (summary table)

	Sizewell 'B'		Coal	
	CEGB costs	Alternative costs	CEGB costs	Alternative costs
Electricity prices (% change)	+ 1.2	+ 0.5	+ 0.5	+ 0.6
National income (£m, March 1982 prices)	− 180	− 138	− 254	− 265
Employment	−9600	−7700	−18400	−18500

If either Sizewell 'B' or a coal-fired alternative[2] were deferred, national income and employment would be substantially lower in the mid 1990s than they otherwise would be. The job losses associated with deferring a new coal station are particularly large – around 18,000 in the economy as a whole. Moreover, the losses from deferring either power station are large even on the basis of alternative cost assumptions. The main reason is that both Sizewell 'B' and a new coal-fired station would displace some existing oil-fired stations, which would be retired early or operated at lower load factors. The oil saved by the CEGB would be released for export, benefiting the balance of payments and enabling an expansion of national income.

Our second recommendation is therefore that if a new coal station is built there is a case for building it early, ahead of demand.

The information available means that this second recommendation must be less unequivocal than the first (that a coal station is preferable to a PWR). Although a new coal station would certainly benefit the UK economy by releasing oil for export, we are unable to assess whether alternative investments would lead to greater benefits. If the ordering of a new power station were deferred until it could be justified on capacity grounds, funds would be released for other CEGB investments or (*via* the Exchequer) for investment in other industries. At least three options, of varying degrees of relevance to the electricity supply industry, should be considered.

One is to achieve the same savings of oil that would be made possible by a new coal station by converting existing oil-fired capacity to burn coal.[3] Since the turbine generators and ancillary plant would not require major modification, this may be a cheaper way of releasing oil. Moreover, since the direct benefit to the economy of higher oil exports is considerable, investment by the CEGB in the conversion of oil-fired capacity ought not to be evaluated on purely internal commercial criteria, and government assistance with the capital costs would be justified.

A second option would be to invest additional funds in the coal industry. The lower electricity prices which the CEGB hopes will result from the operation of the Sizewell PWR could be produced by a reduction in the cost of coal, rather than investment in new generating capacity. Sizewell 'B' may cost £2 billion or more, including interest during construction; capital investment in the coal industry is currently only £750 million a year. We calculate that even the largest reduction in electricity prices made possible by Sizewell 'B' could instead be achieved by a 3 per cent reduction in the cost of coal to the CEGB. Such a reduction in coal prices, by higher investment and productivity, might be attained at less cost than the construction of a new power station.

A third possible use of funds released by deferring the construction of a new power station would be additional investment in manufacturing industry. Infusions of some £2.4 billion of public money into British Leyland since 1975 have by saving the company had a large positive impact on trade and hence on national income. If it could be argued that, pound for pound, investment in manufacturing firms would benefit the balance of trade more than a new power station, then this could well be the preferable course of action.

Nevertheless, to return to our main point, there is certainly a case for the early construction of a new power station because of the potential release of oil for exports, and if a new power station is built, current evidence on the likely macroeconomic impact indicates that it should be coal-fired.

APPENDIX   *Derivation of the economic consequences of power station construction and operation*

This appendix describes the methods used to calculate the income and employment estimates of Chapters 4 and 5. Three separate stages are necessary to reach the figures for the ultimate impact on national income and employment. The first is the identification and subdivision of CEGB expenditure on the construction or operation of a power station. The methods used in this stage are fully documented in the study, and are not repeated here. The second stage involves the calculation of the direct effects of CEGB expenditure on other UK industries and on imports. CEGB expenditure is allocated to individual industries, and the second and subsequent rounds of expenditure of these and other industries are traced to the point where all of the original (CEGB) expenditure has leaked into incomes, imports or taxes. The portion of this expenditure which becomes value-added in UK industries is used to calculate the direct impact on employment. In the third stage, the final effects are calculated. These include the multiplier effects of the expenditure which flows from the increases in personal incomes, and the fiscal policy adjustments made by the government to restore the balance of payments and the public sector deficit to what they would otherwise have been. Stages two and three are described in greater detail below.

Economic projection to 2030

The methods outlined in this appendix rest on estimates of critical aspects of the economy over the next fifty years. Since it is impossible to make justifiable projections over such a long period, we have not attempted to do so. For the period up to 1990 we have used the projection 'reflation with devaluation' published in the *Cambridge Economic Policy Review*, Vol. 8, No. 1. This assumes that government policies are such as to leave unemployment in 1990 close to current levels. Beyond 1990 we have merely assumed a rate of growth for GDP of 1.0 per cent p.a. in order to reflect the CEGB's central economic scenario for the future.

The other aspects of our economic context are shown in Table A1. They are derived in such a way as to be consistent both with a 1.0 per cent p.a. growth in GDP and with one another, and also to avoid implausibly large changes in either unemployment or the current account of the balance of payments. The figures for GDP per employee, value-added per employee in manufacturing, and exports of manufactures are extrapolations of the actual and projected relationships between these and GDP over the 1970s and 1980s.

The current account is assumed to be broadly in balance after 1990. Imports of manufactures after 1990 are calculated as a balancing item between a current account balance and exports of manufactures on the one hand and projected net trade in non-manufactures on the the other. The ratio of food and raw material trade to GDP is assumed to continue its trend decline, while assumptions on fuel exports and imports are drawn from the CEGB's own energy projections. Trade in services are also assumed to continue its trend relative to GDP. It should be noted that calculating imports of manufactures in this way implies a slowdown in the growth of import penetration for manufactures after 1990. Home demand for manufactures is calculated by assuming that the share of manufactures in total domestic expenditure is unchanged. The trade assumptions in Table A1 would be consistent with a growth in world trade of about 7.0 per cent p.a.

Direct effects of CEGB spending

The construction and operation of a power station involves a large volume of purchases from a range of firms both in the UK and abroad. These firms in turn buy components, materials and services from other firms, again in the UK or abroad, and so on. We need to know how much of this expenditure ends up as value-added of UK companies (and hence as incomes within the UK), rather than as imports or taxes. To calculate this we have used coefficients for value-added, bought-in services and materials from UK suppliers, imports, and indirect taxes, in appropriate industries from the latest available UK input-output tables.[1]

Table A1 The economic framework used in assessing the consequences of power station construction and operation

(March 1982 prices)

	1982	1988	1996	2003	2020	Growth % 1990–2020
GDP (£ bn)	271	303	334	358	424	1.0
GDP per employee (£th)	12.3	13.7	15.1	16.1	19.1	1.0
Value-added per employee in manufacturing (£th)	10.5	13.3	16.2	18.6	26.0	2.0
Exports of manu- factures (£bn)	38.4	45.5	56.6	65.6	98.2	2.4
Imports of manu- factures (£bn)	34.2	53.9	70.3	75.8	81.4	1.3
Ratio of imports to home demand for manufactures (%)	27.0	38.1	45.0	48.6	61.5	—
Import penetration for all goods and services (%)	32.1	34.9	37.7	39.0	41.0	—
Balance of payments (£ bn)	4.0	−4.2	0	0	0	—
NCB productivity (tonnes per man year)						
Fast growth (2.7% p.a.)	494[a]	—	748	907	1450	
Slow growth (2.0% p.a.)	494[a]	—	665	764	1069	
CEGB Revenue (£bn)[b]	4.76	6.89	9.00	11.37	16.98	

[a] NCB *Annual Report and Accounts*, 1981/2.
[b] Calculated from CEGB estimates of demand for electricity (*P5*, tables 93 and 50), and of the electricity price with a 'no new nuclear' background (*P4*, Fig. 3).

These coefficients are however amended in several ways. Firstly, they are altered to reflect projected increases in import penetration shown in Table A.1. The coefficients for value-added, bought-in services and materials, and indirect taxes are consequently reduced *pro rata* with the increase in the import coefficient. Secondly, as outlined below, they are altered when individual industry circumstances make it inappropriate to use the input-output coefficients. Second and subsequent rounds of spending are captured in a separate category of 'other industries', comprising all industries supplying bought-in materials and services to those companies which directly supply the CEGB. Allowance is made for transactions within this category of 'other industries' through increasing the value-added, import and tax coefficients (V, M and T respectively) by a multiplier $1/(V + M + T)$. These methods implicitly ignore the small feedback effects into increased demand for electricity and for the products of the industries directly supplying the CEGB.

The employment implications of CEGB expenditure are calculated from the value-added figures obtained using the methodology described above. Figures for value-added per employee were constructed from 1974 data on sectoral value-added in current terms[2] and 1974 data on sectoral employment,[3] adjusted to March 1982 values using earnings indices.[4] Employment figures presented also include a further adjustment to account for productivity growth over time, as described in Table A1.

Construction phase

Total expenditure on power station construction can be allocated to four principal industrial sectors:[5] industrial plant and steelwork, electrical machinery, construction, and NNC and CEGB engineering. The representative year in the middle of the construction phase was taken to be 1988 and in the first three sectors the import proportion was increased by 109 per cent above 1974 levels in line with projected rises in import penetration.[6] There is no relevant

Table A2 Expenditure coefficients for industries supplying the CEGB during the construction phase

	Imports	Value-added	Bought-in goods and services from UK suppliers	Net taxes	Value-added per employee (March 1982 prices)
	%	%	%	%	£
Mechanical engineering	12.8	34.8	51.3	1.1	11300
Electrical engineering	25.6	39.7	33.6	1.1	9500
Construction	10.7	55.5	32.7	1.1	—
NNC and CEGB engineering	5.0	79.0	14.9	1.1	17300
Other industries	19.5	76.6	—	3.9	13400

Notes: Employment in construction was estimated directly from CEGB (1982), *Statement of Case to the Sizewell 'B' Power Station Public Inquiry, Vol. 1 Appendix H*, Fig. 30.21.

In NNC and CEGB engineering it was assumed that average wages were approximately twice the national average.

category in the input-output tables for NNC and CEGB engineering and the proportions shown in Table A2 were assumed so as to reflect the high value-added and low import content of a sector dominated by technologically-advanced, highly skilled design work.[7] As outlined above, the 'other industries' category comprises the industries supplying most of the bought-in component of the sectors listed above.[8] Table A2 shows the complete set of coefficients used for the construction phase.

Operating phase

During the operating phase there are three sectors to be considered:[9] the coal industry, the railways, and nuclear fuel production, plus the 'other industries' supplying these three. Because the period of operation stretches nearly fifty years into the future more assumptions are required than in the construction phase and it must be emphasised that the results are only illustrative.

For the coal industry two sets of assumptions are employed: one which we regard as consistent with the CEGB's views and which involves the industry becoming profitable by the mid 1990's, and another consistent with our alternative coal price assumption involving the continuation of government subsidies to the industry. In both cases we assume a decline over time in the ratio of value-added to bought-in goods and services, reflecting the increasing mechanisation and hence growth of productivity in the industry. The employment effects of power station operation are estimated directly from the volume of coal used and saved rather than from the value-added. In the tables based on CEGB costs, productivity is assumed to grow at 2.0 per cent p.a. after 1982. In the 'alternative cost' scenario, productivity is assumed to rise faster, at 2.7 per cent p.a., to be consistent with the assumption of lower coal prices. In each of the comparisons in this study, the operation of Sizewell 'B' instead of alternative power stations results in reduced demand for coal. The marginal source of coal is assumed to be low productivity pits, which we assume have an output per person equal to only half the industry average in Table A1. Hence reductions in demand for coal result in relatively large reductions in mining employment. The level of subsidy in low productivity pits in the alternative cost scenario is assumed to be 30 per cent.

For the railways we assume that any reduction in coal-carrying business reduces revenue more than it reduces costs. Thus reductions in revenue lead to increases in subsidy. The value-added and bought-in components, derived from the input-output tables, are consequently reduced in proportion to their shares in total inputs to reflect the change in subsidies. In estimating the employment effect of a change in value-added on the railways a further adjustment is made, reducing the employment figure by two-thirds to reflect the fact that the marginal change in employment on the rail network resulting from a change in the coal-carrying business is likely to be less than the average employment per unit of value-added.

For the nuclear fuel industry we assume the import content to be approximately the uranium ore component plus half the enrichment component of the fuel cycle during the 1990s.[10] Assuming that net taxes on expenditure are the same as in the other two sectors, the remaining total inputs are divided equally between value-added and bought-in goods and services.

As in the construction phase, the second-round spending effects are included in the 'other industries' category. This comprises the major industries supplying most of the bought-in component of the three sectors.[11] Table A3 shows the coefficients used for the three representative years during the operating phase.

Final effects of power station expenditure

Each of the options evaluated in this study involves a transfer of income between consumers and taxpayers on the one hand, and those dependent on the electricity industry and its direct and indirect suppliers on the other. During the construction phase the transfer is from taxpayers, who are assumed to reduce their consumption, to the CEGB which increases its investment by building a power station. During the operating phase, a decision to operate one power station rather than another involves gains (or losses) in the incomes of those working

Table A3 Expenditure coefficients for industries supplying the CEGB during the operating phase

	year	Imports %	Value-added %	Bought-in goods and services from UK suppliers %	Net taxes %	Value-added per employee (March 1982 prices) £
Coal industry						
CEGB costs	1996	2	47	50	1	—
	2003	2	45	52	1	—
	2020	2	32	65	1	—
Alternative	1996	2	62	66	−30	—
costs	2003	2	60	68	−30	—
	2020	2	53	75	−30	—
Railways	1996	0	25	35	40	13500
	2003	0	25	35	40	14400
	2020	0	25	35	40	17100
Nuclear fuel	1996	60	20	60	1	19200
production	2003	64	19	64	1	20400
	2020	73	15	73	1	24200
Other industries	1996	40	57	0	3	14700
	2003	45	52	0	3	15700
	2020	55	42	0	3	18600
CEGB	1996	n.a.	n.a.	n.a.	n.a.	16100
	2003	n.a.	n.a.	n.a.	n.a.	17200
	2020	n.a.	n.a.	n.a.	n.a.	20300

for the CEGB and its suppliers, and equal gains or losses to consumers in the form of changes in electricity prices and taxes.

These transfers of income need not however be symmetric, in the sense that what is lost by one group is gained by another. It is this 'asymmetry' which causes total national income to rise or fall. There are three causes of asymmetry. Firstly, the import content of expenditure by the CEGB and its direct and indirect suppliers need not be the same as that of the equivalent expenditure by consumers and taxpayers. Since different amounts of each type of expenditure can leak abroad as imports, the remaining expenditure, which is received as incomes within the UK, can also differ. The subsequent spending of these incomes by their recipients magnifies the differences. To avoid unnecessary complication we assume that any changes in incomes are fully reflected in expenditure.

The second cause of asymmetry is the change in the price of electricity, which results in higher or lower industrial costs. We assume that cost changes are fully reflected in prices, and that the value of imports and exports alters in consequence. Increases in exports lead to a rise in domestic income while increases in imports cause domestic incomes to fall.

Finally, asymmetry in the gains and losses of income is caused by changes in government tax and expenditure policies themselves, resulting from the changes already discussed. We assume that the government acts to restore any change in balance of payments or in the public sector deficit resulting from the construction or operation of Sizewell 'B'. The assumption that domestic income changes are fully reflected in expenditure means that changes in the balance of payments will be equal to changes in the public sector deficit. We assume that domestic expenditure is raised (or lowered) by changes in taxation or government spending up to the point where the change in imports induced by the expenditure offsets the imbalances in trade and government finances resulting from power station expenditure.

In detail, the calculations are performed as follows.

(a) Imports resulting from the expenditure of the CEGB and its direct and indirect suppliers are estimated by the methods outlined in previous sections.

(b) The changes in trade resulting from higher or lower electricity prices are added to the imports from (a) above. Changes in electricity prices are estimated from the changes in operating costs divided by the projected total revenue of the CEGB. Estimates of the latter are given in Table A1. In the construction phase the relevant costs comprise only interest during construction which has been calculated using a 2.5 per cent real rate of interest. The effect of changes in electricity prices on the balance of trade is calculated as an average for manufacturing industry. Electricity costs are currently around 1.0 per cent of total costs in manufacturing and this proportion is assumed to hold throughout the study period. Since variable costs are on average some three-quarters of total costs, electricity costs are 1.33 per cent of variable costs, and a 1.0 per cent increase in the former is thus assumed to increase total costs of manufacturing, including overheads, by 0.0133 per cent. We have assumed an average price elasticity of volume demand for exports and imports of -2, giving an elasticity of unity for demand in value terms. Hence a 1.0 per cent increase in prices results in a 1.0 per cent increase in exports of manufactures plus a 1.0 per cent decrease in manufactured imports. The magnitude of exports and imports of manufactures are given in Table A1. Finally the import content of the net change in trade must be subtracted. This is defined as $C(X-M)$ where X and M are the values of exports and imports respectively, and C is the ratio of imports to home demand for manufactures given in Table A1.

(c) The aggregate direct impact of CEGB expenditure on trade is given by the sum of (a) and (b). An offsetting change in trade is necessary to restore the trade balance to what it would otherwise have been. The change in domestic expenditure necessary to raise or lower imports by enough to restore the trade balance is calculated by dividing the necessary offsetting change in trade by the appropriate import propensity from row 7 of Table A1. This change in domestic expenditure is itself a composite of three effects which do not need to be separately identified. The first is the change in expenditure by consumers and taxpayers consequent on the tax and price changes resulting from building or operating Sizewell 'B'. The second is the difference in the multiplier effects of this spending compared with the direct expenditure of the CEGB and its suppliers. Finally, there is the adjustment in tax receipts or spending of the government. This change in domestic expenditure must be subtracted from the expenditure of the CEGB to obtain the net change in national expenditure. This in turn is equal to the net change in national income.

Employment

Estimates of the final changes in employment are calculated as the sum of three separate components. First is the employment generated by the expenditure of the CEGB and its direct and indirect suppliers. This is obtained as described in previous sections. Second is the employment change in manufacturing due to the alterations in output occasioned by the gains or losses in trade which result from higher or lower electricity prices. Employment in this case is given by the change in income divided by value-added per employee in manufacturing from Table A1. Finally the employment resulting from changes in indirect income must be added. Indirect income is defined as the net change in national income *less* the direct income changes and the income changes in manufacturing resulting from higher or lower electricity prices. Changes in indirect income divided by GDP per employee (from Table A1) give the consequent employment change.

Notes

Notes to Chapter 1

1 The current level of generating capacity is 55GW.
2 See CEGB, *Proof of Evidence, P4* (1982): F. P. Jenkin, *On: The Need for Sizewell 'B'*.
3 T. F. Cripps and W. A. H. Godley (1978), *The planning of telecommunications in the United Kingdom*, Department of Applied Economics, Cambridge.
4 R. E. Rowthorn and T. Ward (1979), 'How to run a company and run down an economy: the effects of closing down steel-making in Corby', *Cambridge Journal of Economics*, 3, pp. 327–340.
5 House of Commons Select Committee on Energy, session 1980–81, *The Government's statement on the new nuclear power programme*, HMSO.
6 The Monopolies and Mergers Commission (1981), *The Central Electricity Generating Board*, HMSO.
7 See CEGB (1983), *Analysis of Generation Costs*, Illustration Three; and G. MacKerron (1982), research report No. 6.
8 CEGB, *op. cit.*, Illustration III; and G. MacKerron, *op. cit.*, pp. 16–22.
9 The CEGB now consider it likely that a number of the Magnox stations will operate for 30 rather than 25 years (CEGB, *op. cit.*, Appendix F). This implies that on central forecasts of electricity demand growth a shortfall in generating capacity is not now expected to occur until after 1996.

Notes to Chapter 2

1 MWso = megawatts sent out (i.e. total power generated less that used for station auxiliary purposes).
2 C. Komanoff (1981), *Power plant cost escalation*. Komanoff Energy Associates, New York. C. Komanoff (1982), 'The Westinghouse PWR in the United States: cost and performance history' paper presented at the Polytechnic of the South Bank, October 1982.
3 Excludes the cost of initial fuel.
4 A power station's load factor is the actual electricity generated during a year expressed as a percentage of the electricity that would be generated if the station were run at full power for the whole year.
5 See the Electricity Consumers' Council (1983), *ECC/P/3*: S. Thomas, *Plant Related Variables: Plant Operation*, p. 17.
6 See the Electricity Consumers' Council (1983), *Statement of Case to the Sizewell 'B' Power Station Public Inquiry*, pp. 11–20.
7 Excluding initial fuel. This is within the range of uncertainty considered reasonable by G. MacKerron. See the Electricity Consumers' Council (1983), *ECC/P/2*: G. MacKerron, *Construction Times and Costs*, p. 39.
8 CEGB, *Proof of Evidence, P6* (1982): P. R. Hughes, *On: Fossil Fuel Supplies*.
9 These include the cost of materials, power, plant hire, compensation for surface damage and rates.
10 A. Jones and K. Woolley (1983), *The evaluation of coal-winning technology*, Technical Change Centre, London.
11 See CEGB, *Proof of Evidence, P5* (1982): C. H. Davies, *On: Scenarios and Electricity Demand*.
12 Calculated from CEGB, *Proof of Evidence, P4*, Table 24, Scenario C.
13 Monopolies and Mergers Commission, *op. cit.*, Table 4.5. The CEGB currently expect excess capacity, as opposed to the planning margin, to peak at 41.5% in 1986/87 and then fall to 34% by 1989/90.
14 An offsetting factor would be the tendency for the costs of all types of power stations to increase over time in real terms in the UK (see the Electricity Consumers' Council (1983), *Statement of Case to the Sizewell 'B' Power Station Public Inquiry*, p. 27).

15 'Retro-fitting' of additional safety features to reactors in operation or under construction has occurred in the United States, notably after the Three Mile Island accident.
16 The same point is made by the Electricity Consumers' Council, *op. cit.*, p. 26.
17 CEGB, *Proof of Evidence, P4*, Table 26c.
18 CEGB, *Proof of Evidence, P4*, Table 29.
19 CEGB (1982), *Statement of Case to the Sizewell 'B' Power Station Public Inquiry*, Vol. 1, para 7.20, p. 26.
20 CEGB, *Proof of Evidence, P4*, Table 26c: time-profile of capital spending on coal stations from private correspondence with CEGB Finance Dept.

Notes to Chapter 3
1 This is broadly in line with the definition used by Fishwick in his study for the CEGB of the employment effects of Sizewell 'B' (see F. Fishwick (1982), *The Effects on Employment within the United Kingdom of the construction of Sizewell 'B' PWR station*, Cranfield Institute of Management, Bedford).
2 *Cambridge Economic Policy Review*, Vol. 8 No. 1, Table A11, p. 57.

Notes to Chapter 4
1 For Sizewell 'B' these allowances for possible overruns in construction costs have been allocated in line with the proportions indicated in F. Fishwick, *op. cit.*, p. 12. For the coal-fired station they have been allocated in proportion to its share in total costs.
2 Specifically, expenditure on the turbine generators, other mechanical plant and other electrical plant have been left constant whilst the other categories have each been increased by 65 per cent.
3 Evidence given before the House of Commons Select Committee on Energy, reported in *The Guardian*, 3 December 1982.
4 See F. Fishwick, *op. cit.*, Appendix 5, p. 45.
5 These calculations have been undertaken for a year during the middle of the construction phase, taken to be 1988. Import propensities and value-added per employee have been adjusted accordingly, in line with forecasts published in the *Cambridge Economic Policy Review*. See Appendix for details.
6 Based on estimates published in the *Cambridge Economic Policy Review*. Vol. 8 no. 1.
7 Including net imports arising from the higher level of electricity prices necessary to finance interest during construction.
8 As long as marginal changes in private sector income are all spent rather than saved, any improvement in the balance of payments is by definition equal to the reduction in the government's financial deficit.

Notes to Chapter 5
1 The estimate of a 35-year lifetime for Sizewell 'B' is criticised by the Electricity Consumers' Council, *op. cit.*, p. 19.
2 Based on the regional distribution of the least profitable pits, reported in *The Guardian*, 24 November 1982.
3 Derived from information in CEGB, *Proof of Evidence, P4*, Figure 16 and Table 25c.

Notes to Chapter 6
1 Calculated using CEGB central forecasts of electricity demand, CEGB forecasts of electricity prices (for the 'no new nuclear' background) and marginal changes in CEGB costs resulting from the construction and operation of Sizewell 'B'.
2 The figures are again for a 'notional' station with the same output as Sizewell 'B'.
3 The CEGB do not seriously address this issue. See Electricity Consumers' Council, *op. cit.*, p. 11.

Notes to the Appendix

1 Business Monitor (1981), *Input-Output Tables for the United Kingdom*, PA1004 HMSO, Table D.

2 See Business Monitor (1981), *op. cit.*, Table D, pp. 37–41.

3 See Business Monitor (1978), *Report on the Census of Production: Summary Tables*, PA1002, HMSO, Table 2, pp. 42–58; and Central Statistical Office (1978), *Monthly Digest of Statistics*, No. 391 July, HMSO, Table 3.2, p. 17.

4 See Central Statistical Office (1978), *op. cit.*, Tables 17.4 and 17.5, pp. 151–154; and Central Statistical Office (1982), *Monthly Digest of Statistics*, No. 439 July, HMSO, Table 18.4, pp. 145–146.

5 This breakdown was determined by information supplied by the CEGB. Access to a more detailed analysis of capital expenditure would have allowed a more precise industrial classification.

6 Projected figures for net manufacturing output, imports and exports in 1990 were derived by linear interpolation of figures for 1986 and 1990 given in the *Cambridge Economic Policy Review* (1982), Vol. 8 No. 1, Tables A.6, p. 52 and A.7, p. 53. Figures for net manufacturing output were adjusted to a gross basis using the ratio for 1979 derived from Business Monitor (1979), *Report on the Census of Production: Summary Tables*, PA 1002 HMSO, Table 1, p. 9.

7 The proportion attributed to net taxes on expenditure was assumed to be the same as that calculated for the three principal supplying industries.

8 Sixteen industries are included: stone, slate, chalk, sand etc. extraction; other iron and steel; other non-ferrous metals; other non-electrical machinery; industrial plant and steelwork; electrical machinery; insulated wire and cables; other metal goods; bricks, fireclay and refractory goods; pottery and glass; other building materials etc.; timber and miscellaneous wood manufactures; plastic products n.e.s.; construction; distributive trades; and other services.

9 A fourth sector, the oil industry, is considered in the text but since we assume that any oil freed from use in power stations will be exported, this has an effect on the balance of payments and hence the level of activity in the economy, but no 'direct' impact on value-added and employment in the industry.

10 See CEGB, *Proof of Evidence*, P9 (1982): J. K. Wright, *On: The Nuclear Fuel Cycle*, Tables 4–7.

11 Fourteen industries are included: gas, electricity, mineral oil refining, lubricating oils and greases, general chemicals, other iron and steel, other non-electrical machinery, other vehicles, other metal goods, other building materials etc., timber and miscellaneous wood manufactures, construction, distributive trade and other services.